EXPLODING DIGITS

Amazing Patterns in Decimal Fractions:

$$^8/_{11} = .72\ 72\ 72\ldots \qquad ^8/_{41} = .19512\ 19512\ldots$$

$$^8/_{23} = .3478260869565217391304$$
$$3478260869565217391304\ldots$$

William B. Martin

ISBN-13: 978-1517361167
ISBN-10: 1517361168

iii

DEDICATION

I dedicate this book to the students of Pima Community College,
Tucson, Arizona

Contents

The author is indebted to his wife, Julia Gousseva, the website www.maths.surrey.ac.uk, and the software program, Mathematica.

Introduction

This book is about the decimal expansion of fractions. The best way to start is to throw a decimal expansion at you:

$$1/13 = .076923\ 076923\ 076923\ \dots \text{ (forever!)}$$

We say 1/13 has a 6 digit repeating pattern. More simply:

$$1/13 = .[076923]$$

Note: Square brackets will be used throughout the book to indicate the set of digits which repeat. This is the same as the more usual line over the digits.

Let's continue with the reciprocals of some more primes. The table below summarizes the behavior of some prime reciprocals.

The Decimal Version of the Reciprocals of Some Primes

Reciprocal	Number of Digits in Repeating Pattern	Decimal Expansion
1/3	1	.[3]
1/5	Terminates	.2
1/7	6	.[142857]
1/11	2	.[09]
1/17	16	.[0588235294117647]
1/29	28	.[0344827586206896551724137931]
1/31	15	.[032258064516129]
1/41	5	.[02439]

The challenge is to find patterns. For example there is one simple rule about the length of the repeating pattern. Did you notice that no reciprocal of p had a length greater than (p-1)? Sometimes the length is much less as in the case of 1/11 or of 1/41. Are there rules when the length is much less? Yes, as we will see.

Both 1/7 and 1/13 have a 6 digit repeating pattern. Why do these two "prime reciprocals" have equal lengths? Do any other primes share equal lengths? Again, we will see later.

There is much more to decimal expansions than their lengths. It is profitable to look at a whole family of fractions such as:

$$1/13, ,2/13, 3/13,\ldots, 12/13.$$

The decimal expansions of these 12 fractions have many patterns and such families are explored in chapter 4. The reasons for the patterns will be proved using simple number theory introduced in chapter 3.

What about composite denominators? What patterns do their decimal expansions have? They have their own rules and are in chapter 5.

One technical word. We can see the digits repeating with a calculator but only if the pattern is short enough—Impossible with the mammoth 28 digit repeating pattern of 1/29! There is a very nice website for the decimal expansion of fractions that I am indebted too. It is

www.mathssurrey.ac.uk

The reader is strongly encouraged to get familiar with this website. Use it to verify statements made and to explore other fractions you may think of.

A new mathematics book is bound to have errors. I would be grateful for any comments at:

william.martin@pima.edu

1 Three Types of Fractions and Two Elementary Facts

Although the subject of fractions as decimals seems quite limited, there are many fascinating patterns in the repeating digits of non-terminating fractions. For example, here are two fractions with denominator 7. They have a fascinating pattern if you are observant:

$$\frac{1}{7} = .142857\ 142857\ 142857... = .[142857]$$

$$\frac{3}{7} = .428571\ 428571\ 428571... = .[428571]$$

As stated in the introduction, square brackets will used throughout the book to indicate the set of digits which repeat.

Do you see what is happening? Both are infinite repeaters with a 6 digit repeating pattern. But there is much more! Both decimal fractions use the same digits. Still more, the digits are in the same cyclic order! Cyclic, as in a circle, means starting with a different digit but in the same order. I call the six proper fractions with denominator 7, the family n/7. Table 3 at the back of the book collects many facts about different families of fractions. Why these families have these decimal characteristics is the subject of this book.

There is more to decimal expansions than mere digit crunching. Deeper mathematical ideas from number theory (the queen of mathematics) and infinite series will be used. The widespread availability of handheld calculators allows facts to be quickly verified and the reader will have more fun if he or she uses one as the book is read.

Here are three fractions that illustrate the three types of decimal fractions. Find the decimal versions of the three fractions below using a calculator:

$$\frac{11}{16}, \frac{5}{37}, \frac{47}{82},$$

There are three different behaviors illustrated:

1) Decimals that terminate
2) Infinite repeating decimals with a pattern that starts immediately after the decimal point
3) Infinite repeating decimals but the pattern starts sometime after the first decimal place

Did you see the three types?

$\dfrac{11}{16} = .6875$. *It terminates,*

$\dfrac{5}{37} = .[135]$, *and the repeating pattern starts immediately*

after the decimal point.

Finally,
$$\frac{47}{82} = .57[31707].$$

This fraction has a five digit repeating pattern but unlike the second

fraction the pattern starts after two initial digits.

The purpose of this book is to explore fractions with infinite repeating patterns. What determines the length of the patterns? A great way to organize fractions is to look at an entire family of proper fractions with the same denominator. One family is those written as: $n/13$, with
$n = 1,2,3, \ldots ,12$. It turns out that half of the twelve fractions in this family have the same set of repeating digits. The other six all have the same six digits but they are a different set of digits from the other six fractions! In chapter 4 we will organize these 12 fractions into what are called clocks.

Fractions with repeating decimals that do not start at the decimal point have interesting variations on this. They are discussed in chapter 5.

Let's dispose of the simplest type, terminating fractions. Which fractions terminate? The only terminating fractions are those whose denominators factor into powers of 2, 5 or combinations of these two numbers. Examples are:

$$\frac{1}{4} = .25, \qquad \frac{1}{16} = .0625, \qquad \frac{43}{125} = .344,$$

$$\frac{53}{200} = .265, \quad \text{and} \quad \frac{11,433}{4000} = 2.85825$$

Notice that our rule applies to the last fraction which is greater than one, called improper fractions. In this book we will deal only with proper fractions.

Of course, in other bases the story is different. For example, in base 12 many reciprocals terminate. Here are the expansions of some reciprocals of integers in base 12.

$$\frac{1}{2} = .6_{12} \qquad\qquad \frac{1}{3} = .4_{12} \qquad\qquad \frac{1}{4} = .3_{12}$$

$$\frac{1}{5} = [.2497]_{12} \qquad\qquad \frac{1}{6} = .2_{12} \qquad\qquad \frac{1}{7} = [.1\ 8\ 6\ 10\ 3\ 5]_{12}$$

The last example says that one-seventh is a 6 character repeating number with 10 in the 4th position. Places were separated by a space. We will stick mainly to base 10 in this book but some things are true in any base. For example, any fraction with a prime denominator terminates or immediately starts repeating in any base not equal to the prime or a multiple of it. Reasons for some statements will come later. Have a calculator handy, but when the expansion is long or to find the expansion in any base: use the website:

www.maths.surrey.ac.uk

Much of the book's decimal expansions were done using this website.

I finish this chapter with two elementary facts about decimal fractions.

   Fact 1. All non-terminating fractions have a fixed pattern of decimals that repeats forever. Why?

   Think of the long division process. In calculating the decimal version of a/b the only possible remainders come from the integers $\{1,2,3,...,b-1\}$. If you are not sure what I mean by remainders, see the long division example in chapter 3. That example shows that when a previously encountered remainder occurs the remainders and their associated digits in the quotient will repeat forever!

   Fact 2. For any proper fraction of the form:

$$\frac{d_1 d_2 ..... d_n}{10^n - 1}$$

The decimal expansion is the infinite repeating pattern:

$$.[d_1 d_2 d_3 ... d_n]$$

Note: $10^n - 1$ is a string of n nines

   The proof of Fact 2 is tedious. Here is a proof for n equal to two which can then be generalized.

   We will use the well-known infinite geometric series formula. For any r in the interval (-1,1) we have:

$$1 + r + r^2 + ... = \frac{1}{1 - r} \quad (1)$$

   Suppose we have the two digits ab[1] divided by 99. We will prove that the decimal version is .[ab]:

---

   [1] We are really speaking of two different digits a and b. If they were equal it would be a one digit repeater.

$$\frac{ab}{99} = \frac{ab}{100} \frac{1}{99/100} = \frac{ab}{100} \frac{1}{1 - 1/100}$$

$$= \frac{ab}{100}\left(1 + \frac{1}{100} + \frac{1}{100^2} + ...\right)$$

$$= \frac{ab}{100} + \frac{ab}{100^4} + \frac{ab}{100^6} + ...$$

$$= .ababab... = .[ab].$$

The extension to a denominator with any number of nine's can be done using mathematical induction.

If the numerator has less than n digits then the repeating pattern has leading zeros:

$$\frac{813}{99999} = [.00813]$$

Here is a fraction that can be reduced;

$$\frac{135}{999} = \frac{5}{37} = .[135]$$

In fact, any fraction of the form n/37 is a three digit repeater. The reason is an important theorem in the next chapter.

Chapter 1 Exercises

For exercises 1 through 4 , consider the 100 fractions:

$$1/2, 1/3, 1/4, ..., 1/101.$$

Note: The following website will be very useful in answering the exercises.

www.mathssurrey.ac.uk

1. Fourteen reciprocals terminate. Which ones?

2. How many fractions are immediate repeaters ? Which ones?
Hint: In addition to the reciprocals of primes there are also the reciprocals of some composite denominators. More than 35 in all!

3. Which fraction has the longest repeating pattern?

4. Using Table 1 in the back of the book find the biggest reciprocal (smallest prime denominator) that is an immediate repeater and has a repeating pattern of length:
   a) 3   b) 4   c) 5

2 The Length of Repeating Patterns

It is easy to predict the length of the repeating pattern of any reduced fraction. Find the prime factorization of the denominator. Then consider the infinite set of integers

$$\{(10^n - 1) \ for \ n = 1,2,3,...\}$$

Note: These are integers consisting of n nines. We use the following theorem.

Length of Period Theorem (LPT): Consider a reduced fraction with denominator d.. If d has any factors of 2 or 5 disregard them. Find the smallest number of the form $(10^n - 1)$ that d, exclusive of 2 and 5, divides into evenly. The length of the repeating pattern equals n. In other words, the repeating pattern equals the number of nines in the smallest string of nines divisible by the denominator after any powers of 2 and 5 are discarded. Here are three examples before a proof:

Example 1:.
$$\frac{7}{(2)(3)(11)} = .1[06]$$

This is a two digit repeater because the factors 3 and 11 <u>both</u> divide into the number consisting of two nines, 99.

Example 2:
$$\frac{19}{(2^2)(5)(3^3)} = .03[518]$$

This is a three digit repeater because 3^3 divides into 999 and no smaller string of nine's

Example 3:
$$\frac{97}{(2)(5^3)(3^2)(101)} = .003[8415]$$

This is a four digit repeater because both 3^2 and 101 divide into the four string, $9999 = (10^4 - 1)$ and no smaller string of nine's.

Table 2 in the back of the book lists factorizations of the all-important strings of nines and has another example. Here is a proof of the Length of Period Theorem.

$Let \dfrac{n}{d}$ be a reduced fraction such that d has no factors of

2 or 5.

(We will see later that factors of 2 or 5 merely delay the start of the repeating pattern). Let s be the length of the smallest string of nine's that is divisible by d.[2] Then for some integer k:

$$dk = 10^s - 1$$

We work with our fraction thus:

$$\frac{n}{d} = \frac{nk}{dk} = \frac{nk}{10^s - 1}$$

But the last version of the fraction has an s-long repeating pattern of digits by Fact 2 of the last chapter.

Let's explore how to use the factorizations in Table 2. The prime factors of $(10^n - 1)$ give the denominators of fractions that have n length repeating patterns or repeating patterns that divide into n. For example:

$$10^3 - 1 = 999 = (3^3)(37).$$

Thus fractions with denominators that are powers of 3 up to the third power have repeating lengths that divide into 3. Since 3 is only divisible 1 and 3 it means that fractions with denominators of 3, 3^2 or 3^3 have either one or three digit repeating patterns. Check this out!.

In Table 2 I have put in bold font the first occurrence of a factor. The number 37 is in bold in line three. This implies that 37 divides into 999 and no smaller string of nines. Reduced fractions with the sole denominator 37 must have a repeating pattern that is exactly three digits long.

Powers of 3 up to the third power in combination with the factor 37 will also have a three digit patterns. Here is a table and another example to illustrate.

---

[2] Will there always be some string of nines large enough to ensure this for any denominator d? Yes, by the Euler's Generalization of Fermat's Theorem in Chapter 5. .

Denominator	Length of Repeating Pattern	Comment
3	1	3 divides into 9
3^2	1	3^2 divides into 9
3^3	3	3^3 divides into 999
37	3	37 divides into 999
3×37, $3^2\times37$, or $3^3\times37$	3	All of these divide into 999

Another example.
$$10^4 - 1 = 9999 = 3^2 \times 11 \times 101$$

The prime factorization 9999 indicates that fractions with denominators of 3, 3^2, 11 or 101 (or combinations of them) have repeating patterns with lengths that divide into 4. The possible decimal length pattern of fractions with these denominators are 1,2 or 4. Try some fractions with these denominators!

The next chapter will use a beautiful connection with number theory to see why prime denominators are immediate repeaters.

Chapter 2 Exercises

Use Table 2 to answer exercises 1 through 3.

1. What size decimal pattern will the following fractions have? State the decimal pattern in exercise 1 and explain your answers using Table 2 in exercise 2..
a)19/41 b) 19/(2*41) c) 19/(2³.*5*41)

2. a) $2/3^2$, b) $2/3^3$, c) $2/3^4$ d) $2/(3^4*11)$ Also explain why in each case.

3. In exercise 1b and 1c what was the effect of the 2's and 5's on the:
a) length of the decimal pattern?
b) Start of the decimal pattern?

3 Applying Modular Arithmetic to Decimal Expansions

To understand the decimal version of fractions we need a small dose of number theory. This beautiful subject will pay big dividends in understanding the decimal expansions of all fractions.

We start with a definition. Two numbers are <u>relatively prime</u> if they don't have any common factors. Thus 5 and 8 are relatively prime but 6 and 8 are not, having a common factor of 2. The symbol (a,b) denotes the greatest common factor of the integers a and b. So:

$$(5,8) = 1 \text{ and } (6,8) = 2.$$

Thus a and b are relatively prime only if (a,b) = 1. Suppose b and p are integers with p equal to any prime. Then (b,p) equals 1 for any number b less than p. For an integer b greater than p, (b,p) = 1 or (b,p) = p if b is a multiple of p.

Finally, we introduce the "mod" symbol of modular arithmetic with an example:

$$17 = 2(\text{mod } 5)$$

We say,"17 equals 2 mod 5". This statement means that when 17 is divided by 5, the remainder is 2. Here are two equivalent statements;

1. The difference of 17 and 2 is divisible by 5.

2. The numbers 17 and 2 both have remainder 2 when divided by 5.

The key to understanding the patterns in repeating fractions is the remainders that occur in the long division process. Consider the conversion of 25/37 to a decimal:

```
          .675
 37 | 25.000
      222
      280
      259
      210
      185
       25(matches the numerator)
```

Pay particular attention to the three remainders in bold font: **28, 21** and **25**. The point is that the last remainder 25, matches the 25 in the numerator of the fraction. We know that bringing down the next zero, (really multiplying by 10) will result in the remainders 28,21 and 25 again. And the next three decimals will be 6,7, and 5, the same as the first three. The remainders <u>dictate</u> the decimal expansion. Since the three remainders, 28,21,25… will repeat forever, the digits 6,7,5 will repeat forever in that order. We can say:

<u>When a remainder repeats a pattern of remainders is set which will continue forever.</u>

There is an important way to see what the remainders will be using modular arithmetic. Returning to our example, the product of successively higher powers of 10 multiplied with 25 mod 37 give the following results mod 37:

$$(10)\ (25) = 28(\text{mod } 37)$$
$$(10^2)\ (25) = 21(\text{mod } 37)$$
$$(10^3)\ (25) = 25(\text{mod } 37)$$

We see the numbers 28,21 and 25, matching the remainders, 28, 21 and then 25 and in the same order as we saw in the long division process.

We can now see the significance for decimal expansions of an important theorem from number theory. It is called Fermat's Theorem[3]:

<u>" Suppose n is a number and p is a prime number. If n and p are relatively prime then</u>

---

[3] Or sometimes, "Fermat's Little Theorem".

$$n^{p-1} = 1(mod\ p)$$

Since our subject is the decimal expansion of fractions, we will use n equal to 10. In this case the theorem says p can be any prime except 2 or 5. This still leaves infinitely many primes of course!

It will turn out that we are interested in the <u>smallest</u> power s, such that 10^s is equal to one (mod p). <u>The reason for this interest is that this smallest power, now designated s, is equal to the length of the repeating pattern in the decimal expansion of any fraction a/p where a is any numerator !</u> The reader may wonder, "Why the emphasis on a power equal to <u>one</u>?". The reason for this emphasis will become clear for reciprocals of primes soon.

A corollary to Fermat's theorem that is important for decimal expansions is that the smallest power s either equals (p-1) or divides evenly into (p-1). This is consistent with the important Length of Prime Theorem (LPT) in the last chapter. A special case of that theorem is for fractions with prime denominator, p[4]. As usual p does not equal 2 or 5. For the reciprocal 1/p, LPT says that the length of the repeating pattern is the smallest s such that $(10^s - 1)$ is divisible by p. While Fermat's theorem says that for n =10:

$10^{p-1} = 1(mod\ p)$
Subtracting 1 from both sides implies

$10^{p-1} - 1 = 0(mod\ p)$

The left side is a (p- 1) long string of nines. Since this integer is equal to zero mod p, it is divisible by p. LPT says find the smallest string of nines divisible by the denominator. The smallest s-string gives the length of the repeating pattern, so LPT is better for finding the length of the repeating pattern. The following table with different primes will show the contrast:

---

[4] Later we see how to work with any denominator d as long
As it has no factors of 2 or 5

Comparison of Fermat's Theorem to Length of Prime Theorem (LPT) for Five Primes

Prime	Fermat's Theorem Says	Goal of LPT is to find the smallest power s of 10 equal to one	Length of Pattern of fractions a/prime (Equals s)
3	$10^{3-1} = 1 (mod\ 3)$	$,10^1 = 1 (mod\ 3)$	^{11}
7	$10^{7-1} = 1 (mod\ 7)$	$10^6 = 1 (mod\ 7)$	6
11	$10^{11-1} = 1 (mod\ 11)$	$10^2 = 1 (mod\ 11)$	2
17	$10^{17-1} = 1 (mod\ 17)$	$10^{16} = 1 (mod 17)$	16
41	$10^{41-1} 1 (mod\ 41)$	$10^5 = 1 (mod 41)$	5

We see that the exponent guaranteed by Fermat's theorem is sometimes the smallest exponent as in the cases for 7 and 17. More frequently, the smallest exponent, s is much less as for the primes 3,11 and 41.

Here is the decimal expansions of the reciprocals of theses primes. Look at the length of the patterns.

$$\frac{1}{3} = [.3] \qquad \frac{1}{7} = [.142857] \qquad \frac{1}{11} = [.09]$$

$$\frac{1}{17} = .[0588235294117647] \quad and \quad \frac{1}{41} = .[02439$$

Why does the smallest power s, such that $10^s = 1(mod(p)$,

match the length of the pattern in the decimal expansion of $\frac{1}{p}$?

The integer $(10^s -1)$ is an s-long string of nines. In chapter 2 we found the smallest such string to find the length of the repeating pattern. We called this procedure the Length of Prime Theorem (LPT) It's often a good idea to prove something a different way. Let's see how we can use mod arithmetic to prove LPT.

The powers of 10 (mod p) exactly match the remainders in the long division process of 1/p. Try expanding 1/7 by long division, no calculator! Note the remainders. For reciprocals (numerator is 1) the remainders will start to repeat when one occurs. Then compute the six powers, 10^x (mod 7) for x equal to 1 through 6. The results will match exactly the remainders in the long division. When a power of 10 equals 1 (mod p) we <u>must</u> start repeating the remainders. The remainders dictate the decimals, so a repeating of the decimal pattern is also starting . We have a theorem:

" For primes p not equal to 2 or 5, the length of the repeating decimal pattern for 1/p equals the smallest power s such that $10^s = 1(\text{mod } p)$".

Moreover, the decimal pattern for 1/p will start immediately after the decimal point and the repeating pattern's length must divide into (p-1) because of Fermat's Theorem. A formal proof is in the next chapter. We call primes whose reciprocals have the maximum possible length, which is (p-1), <u>long primes</u>. Here is a table of the length of the repeating pattern for a few primes.

Prime p	Length of Repeating Pattern of 1/p	Comment
7	6	Long Prime
13	6	6 divides into (13-1)
17	16	Long Prime
19	18	Long Prime
23	22	Long Prime
29	28	Long Prime
31	15	15 divides into (31-1)
37	3	3 divides into (37-1)
41	5	5 divides into (41-1)

The 16 place repeating decimal for 1/17 is way too large to see on a calculator. Use the website mentioned in the introduction for long period repeaters.

We can expand the "smallest power s idea" to other bases not equal to 10.

In the next table are the smallest powers of different bases that give one mod that base. The smallest power s matches the length of the repeating pattern in that base. The repeating version is also given in that base.

1/17 in Different Bases b

Base b	Smallest s, such that $b^s = 1 \pmod{17}$	Expression in base b
2	8	[.00001111]
4	4	[.0033]
13	4	[.0 0 12 3]
33	2	[.1 31]

Note: For bases larger than 10 we separate each place with a space. For example 1/17 is particularly easy in base 33. The entry in that line of the table means

$$\frac{1}{17} = \frac{1}{33} + \frac{31}{33^2} + \frac{1}{33^3} + \frac{31}{33^4} + \dots$$

We will refer to these tables in the next chapter when a whole family of fractions , n/p are organized into clocks. It will be a beautiful organization.

Chapter 3 Exercises

1. The decimal expansion of 1/37 is dictated by the modular values :

$$10 \times 1(Mod\ 37),\ 10^2 \times 1(Mod37),\ and\ 10^3 \times 1,(Mod37)$$

. What are these three modular values and how do they dictate the expansion of 1/37?

2. Consider the analogous case of the decimal expansion of 1/41.
What Mod values are appropriate in this case and how many are needed before repetition will set in?

3. When the numerator is not one, say n/p the appropriate mod values are
Mod(10*n,p), Mod(10^2*n,p)... ,until Mod(10^s*n,p)=n

Set up the mod values needed to find the number of repeating digits in the expansion of 52/101.

4. Find the "decimal" expansion of 1/10 in the following bases. (That www.mathssurrey.ac.uk website can do it.):

2,3,4,5 and 7.

5. For which bases in problem 4 was the expansion 1/10 immediate repeating? Why?

6. For which bases in the range 2 through 17 will the fraction 1/6 be:

a) immediately repeating?   b) repeating but not immediately?

and c) terminating?

7. We saw in chapter 1 that the length of the repeating <u>decimal</u> pattern of 1/p where p is a prime is bounded by the smallest n in (10^n -1) that p divides into. For example 1/7 has a 6 digit repeating pattern. This implies that the smallest n for which 7 divides into (10^n -1) is n =6. Here is a table of the expansions of 1/7 in other

bases. In the last column fill in what the base must divide into as implied by the length of the repeating pattern.

Expansion of 1/7 in other Bases

Base	Length of Repeating Pattern in that Base	Implies that 7 divides into what?— Fill in.
2	3	
3	6	
4	3	
5	6	
6	2	
8	1	
9	3	

4 Prime Denominators and Clocks

There are infinitely many fractions of course, but primes provide a nice way to organize their decimal expansions. In particular, focus on the decimal expansions of a <u>family</u> of proper fractions n/p where p is a prime not equal to 2 or 5. The numerator n will take on the values 1,2,3,... ,(p-1) and will be called the family of fractions, n/p. For such families all the (p-1) fractions in the family have the three characteristics:

1. All decimal versions are immediate repeaters, the pattern starts repeating immediately after the decimal point. So called <u>long primes</u> have a period of length (p-1). All (p-1) fractions of long primes use the same digits in their decimal expansions. Moreover, all (p-1) fractions have digits in the same cyclic order!

2. Suppose p is not a long prime and has a repeating pattern of size s. Then s divides into (p-1), as was stated in the last chapter. All fractions in the family n/p have the same length s repeating pattern. All (p-1) fractions have a pattern that is s digits long. We can say equivalently that each of the (p-1) fractions has a repeating length s that divides into (p-1).

3. When the repeating pattern is of size s, less than (p-1), then the (p-1) fractions of the n/p family splits into (p-1)/s subfamilies. Each subfamily of fractions uses s digits in its expansion. Moreover, each subfamily uses the same digits in the same cyclic order! The behavior of fractions within every subfamily is the same as for fractions of long primes.

Proof of these facts follows. But first let's see a beautiful example.

The 12 member family 1/13, 2/13, 3/13,…, 12/13 are all 6 digit repeaters. In this case s equals 6. Note that 6 divides into (13-1). There will be (13-1)/6 = 2 subfamilies. As stated above, the subfamilies will use the same digits in the same cyclic order. Here are the two subfamilies of n/13.

The Decimal Expansion of the Family n/13

Subfamily using the 6 digits: {0,1,2,3,6,9}	Subfamily using the 6 digits: {1,3,4,5,6,8}
1/13 = .[076923]	2/13 = .[153846]
3/13 =. [230769]	3/13 = .[384615]
4/13 = .[307692]	6/13 =.[461538]
9/13 = .[692307]	7/13 = .[538461]
10/13 = .[769230]	8/13 =.[615384]
12/13 = .[923076]	11/13 = .[846153]

The 12 fractions have been split into two subfamilies based on the digits used. Each subfamily uses the same digits in the same <u>cyclic order.</u> There is much to say about n/13! Each fraction in a subfamily has a different starting digit. The best way to organize and understand all this is to draw two circles or clocks. Directions for making the two clocks follow.

First we will organize the left hand subfamily above. Draw a large circle. Just inside the circumference write the first digit of 1/13, namely 0, at the 12 o'clock position. Continue clockwise inside the circle writing the next 5 digits of 1/13: 7, 6, 9, 2 and 3. Put the 6 digits equally spaced around the circle. Finally, write the 6 fractions just outside the circle, such that each fraction is adjacent to its starting digit. So 1/13 will go right above the 0 on your clock. When done there should be 6 numbers on the inside and just outside the circle next to each digit should be a fraction, corresponding to that fraction's first digit. You can also draw a circular arrow inside the digits and going clockwise indicating the order of the digits. Now the expansion of any fraction in the subfamily is easily read by starting at the digit opposite the selected fraction and going clockwise around the circle, and realizing that the 6 digits repeat forever! Similarly make a second clock for the second sub-family. All 12 fractions have been organized into two clocks and any decimal expansion can be read off easily.

There is yet more to this interesting family, n/13! Notice that the sum of the fractions in each clock is the same, 39/13 = 3. The sum of the digits in each clock is the same, 37. The sum of any two opposite digits is 9 and the sum of any two opposite fractions is 13/13!

So why do families of fractions with prime denominators behave this way?. The key is Modular Arithmetic and Fermat's theorem. For p not equal to 2 or 5 there is a smallest power s such that

$$10^s = 1 (mod\ p) \qquad (1)$$

We can now argue independently of LPT to see why this key value of s applies to all proper fractions n/p and not just to 1/p as we did in the last chapter. There will be some additional benefits in using Modular Arithmetic. But first, why do all fractions n/p where p is a prime not equal to 2 or 5 have the length s of equation (1)?

Suppose the decimal expansion of the proper fraction n/p is desired where p is a prime and n does could be different from 1. It is convenient to consider n as the first remainder. Recall that in the long division process of converting a fraction to a decimal we multiply each remainder by 10 (mod p). Similar to when the numerator is one, the successive remainders are just n times successive powers of 10 (mod p). For a numerator of n, an ordered list of the remainders in the decimal expansion n/p is:

$$n, 10n(mod\ p),\ 10^2 n(mod\ p), 10^3 n(mod\ p)..., 10^s(mod\ (p) \qquad (1)$$

But

$$10^s n\ (mod\ p) = n(mod\ p) \quad because\ 10^s = 1\ (mod p)$$

So the pattern of remainders (and their associated digits) in list (1) has started to repeat. Counting the remainders, there are s (can't count the last one). Thus we have s digits in every expansion of n/p.

Next: Why does any subfamily have the same s digits in the same cyclic order?

Think of a particular subfamily of remainders as:

$$n, 10n(mod\ p),\ 10^2 n(mod\ p),\ 10^3 n\ (mod\ p),..., 10^{s-1} n\ (mod\ p).$$

Let's choose one of these remainders, say

$$10^k n(mod\ p)$$

where k is an integer in the list $\{1,2,3,..., s-1\}$. This remainder has a corresponding fraction in this subfamily, namely

$$\frac{10^k n \ (mod \ p)}{p}$$

The decimal expansion of this fraction will have these remainders <u>in this</u> <u>order:</u>

$10^k n \ (mod \ p)$, $10^{k+1} n (mod_p)$, $10^{k+2} n \ (mod \ p)$,..., $10^{k+s} n \ (mod \ p)$.

But the last remainder above equals the first because

$$10^s = 1 \ (mod \ p).$$

The remainders have started to repeat. We cannot count the very last element in the list (it equals the first). Thus there are $(k+s) - k = s$ remainders. Moreover this list is the same for all fractions in this subfamily. The last bonus from the mod list is that we can see why the order of the remainders in any subfamily is cyclic. This means a list of remainders, as above, can begin with any element of the list An element corresponds to the <u>numerator</u> of one fraction in the subfamily. But then multiplying by successive powers of 10 we are led around in a cycle of remainders which has the same cyclic order as any other fraction in the subfamily. The same is evident for other subfamilies.

Even better for understanding, one can assign an index to each remainder, equal to the power of 10 corresponding to it. There are s indices for each subfamily. A list of remainders (or fractions) in any subfamily can be put into a one to one correspondence with the s indices. The list of remainders of any other fraction in the subfamily corresponds to the same indices via its powers of 10. The only difference is that the indices (and their corresponding remainders) will start with the index corresponding to the numerator of that fraction. The indices (and their remainders) must then cycle around through the same numbers because of the $10^s = 1$ (mod p) property.

Remember that the remainders dictate the digits in the decimal expansion of all fractions. So the digits are the same in one subfamily and likewise the only difference between fractions in a subfamily is the starting digit.

In summary each subfamily of fractions in n/p has period s, where s divides into (p-1). Each subfamily has the same remainders in the same cyclic order and thus each subfamily has the same digits in the same cyclic order.

There is a nice theorem about the sum of the fractions in any subfamily. In other words the sum of the fractions in a clock.

Theorem: For any prime p the sum of the fractions in any subfamily (or clock) is an integer.

Proof. Suppose the decimal period of the fractions with denominator p has length s. Then, as we know:

$$10^s = 1(mod\ p)$$

This means that p divides into

$$10^s - 1$$

This expression is a string of s nines. It can be divided by 9, giving the string of ones:

$$(10^s - 1)/9$$

So p, which is prime, divides into this s--long string of ones.

Suppose n/p is one fraction of the subfamily. We know the remainders in the decimal expansion of n/p are the s numbers:

$$n, 10n(mod\ p),\ 10^2n(mod\ p),\ 10^3n\ (mod\ p),...,\ 10^{s-1}n\ (mod\ p)$$

Now add these numbers and factor out n:

$$n[1 + 10,\ 10^2, 10^3\ ,...,\ + 10^{s-1}]\ (mod\ p)$$

The quantity in brackets is an s--long string of ones. We stated that this number is divisible by p. Therefore the sum of the numerators in any subfamily is divisible by p. This means the sum of fractions in the subfamily is an integer. QED

In the next chapter , exercise 4 there is an interesting contrast for the fractions with the non-prime denominator 27.

I close this chapter with an example. The 40 fractions in the family n/41 are all 5 digit repeaters. There are $(41-1)/5 = 8$ subfamilies. Each subfamily can be arranged into a clock. There will be 8 clocks with 5 fractions in each. One clock has the fractions 1/41, 10/41, 18/41,16/41 and 37/41. Here is a table for this clock with their decimal expansions and their remainders in the order of the long division process.

The Subfamily: 1/41, 10/41,18/41,16/41 and 37/41

Fraction	Decimal	Remainders Including the Numerator Repeated
1/41	[.02439]	1,10,18,16,37 and then 1
10/41	[.24390]	10,18,16,37,1 and then 10
18/41	[.439024]	18,16,37,1,10 and then 18
16/21	[.761904]	16,37,1,10,18 and then 16
37/41	[.90243]	37,1,10,18,16 and then 37

The sum of the fractions in this clock or subfamily is 82/41=2. The sum of the fractions in each clock is an integer, as guaranteed by the last theorem. The reader is invited to make a clock for this subfamily. Why not make all 8 clocks for the 8 subfamilies of n/41?

Chapter 4 Exercises

1. What is the sum of the family of fractions:

$$1/p+2/p+3/p+,\ldots,+(p-1)/p ?$$

2. Let's look at the decimal expansions of two full period primes.
a) In particular find the sum of the digits in the repeating patterns   of:
$$\frac{1}{7} = .[142857]$$

$$\frac{1}{17} = .[058235294117647]$$
and of

b) Can you think of a reason for the sum equal to 27 in one case and 72 in the other?

c) The prime 29 is full period. The sum of the digits in 1/29 is 126. How does this fit the pattern of the part (b) above?

3.) This exercise will explore the 36 digit expansions of the family:

$$1/37,2/37,3/7,\ldots,36/37$$

a) How long are the patterns in these fractions?
b) How many clocks will it take to organize this whole family?
c) Make some of the clocks in this family. Look at the sum of the fractions in some of the clocks. Half of the clocks have one sum and half have another sum. What are the two sums?
d) We saw that the sum of the fractions in each of the 12 clocks was an integer. Suppose we know that the sum of the digits in a three member subfamily is 9 or a multiple of 9. Prove that the sum of the subfamily of fractions is an integer using the digits We can start with a general expression for the three fractions in one clock (or subfamily):

$$.[abc], \text{ and } .[bca] \text{ and } .[cab]$$

Taking the first decimal:

$$.[abc]= abc/999= (100a+10b+c)/999$$

And do the same thing for the other two fractions. Add the three fractions. Show that the sum of the expansions in the subfamily is an integer. An interesting contrast to this is non-prime family n/27 which is discussed in exercise 4 at the end of the next chapter.

5 Composite Denominators

Composite denominators have interesting variations on the behavior of primes denominators. How long are the repeating patterns and do they start immediately after the decimal place? If not how many places after the decimal point does the repeating pattern start? We will use a powerful theorem which is a generalization of Fermat's theorem.

Denominators which consist only 2's and/or 5's terminate as decimals. This chapter deals with non-terminating, composite denominators. These fractions may be grouped into two cases:

<u>Case 1</u> Composite denominators with no factors of 2 or 5.

<u>Case 2</u> Composite denominators with 2's and/or 5's (but also with other prime factors).

Case 1 will be immediate repeaters, meaning the repeating pattern starts right after the decimal point. This is the same as with prime denominators. We will prove this but first we need some number theory. We start with the definition of Euler's function:

The $\varphi(n)$ function:

For a positive integer n, the function $\varphi(n)$ equals the number of positive integers less than n that are relatively prime to n.
By convention, the number one is always counted.
For example

$$\varphi(6) = 2,$$

because there are only two positive numbers less than 6 that are relatively prime to 6. The integers are one and five.

A table of this function for n equal one to nine follows.

The $\varphi(n)$ function for n equal 1 though 9

n	$\varphi(n)$	Integers less than n and relatively prime to n. Note: One is always included. Count them!
1	1	{1]
2	1	{1}
3	2	{1,2}
4	2	{1,3}
5	4	{1,2,3,4}
6	2	{1,5}
7	6	{1,2,3,4,5,6}
8	4	{1,3,5,7}
9	6	{1,2,4,5,7,8}

One obvious fact is that for p a prime:

$$\varphi(p) = p - 1$$

Also very useful is the following case. If r and s are relatively prime then:

$$\varphi(rs) = \varphi(r)\varphi(s)$$

Now we are ready for a new theorem. For composite denominators d we use "Euler's Generalization of Fermat's Theorem":

Suppose c and d are relatively prime integers. Then:

$$c^{\varphi(d)} = 1(mod\ d)$$

For the purposes of decimal expansions we choose c equal to 10. Recall that for Case 1 fractions d has no factors of 2 or 5. So for c equal to 10, d is relatively prime to c and we can use this theorem. Here is an example.

Suppose c equals 10 and d=21. We must find Euler's function for 21:

$$\varphi(21) = \varphi(3)\varphi(7) = (2)(6)=12$$

Then our theorem says

$$10^{\varphi(21)} = 1 \pmod{21}$$

$$or\ 10^{12} = 1 \pmod{21}$$

This is true but, as with prime denominators, we are interested in the lowest power of 10 that gives 1. In this case the lowest power is 6, that is

$$10^6 = 1 \pmod{21}$$

The significance is that all reduced fractions[5] with denominator 21 have repeating decimals that are 6 digits long. Euler's generalization of Fermat's Theorem implies the following for all Case 1 fractions with denominator d :

The number of repeating places equals $\varphi(d)$ or divides into $\varphi(d)$.

A corollary to "Euler's Generalization of Fermat's Theorem" will be assumed:

"For any number d, relatively prime to 10, there is a lowest power s such that:

$10^s = 1 \pmod{d}$ *and any reduced fraction* $^n/_d$ *will have a repeating decimal of length s.*"

As I was writing this I thought briefly of a nice analogy to <u>full period primes</u>:

"*Composite fractions with repeating length $\varphi(d)$ will be called Full Period Composites*"

Then I started to look for some but could not find any! There are no so-called full period composites. Here is a beginning to understanding why. Consider a denominator d in the simple case where of d=st, and where s and t are two different primes. Suppose also that s and t are full period primes.
Then

---

[5] We eliminate fractions such as 14/21 which does not have a composite denominator when it is reduced. From now on always assume n/d is a reduced fraction.

41

$$\varphi(d) = \varphi(st) = \varphi(s)\varphi(t)$$

We will see later that the length of the repeating pattern of any fraction $n/d = n/(st)$ is actually the lowest common multiple (LCM) of the lengths of the repeating patterns of the fractions with denominators s and t. This is inherent in the Length of Period Theorem, LPT. Using this idea the length of the repeating pattern is the LCM of the lengths $1/s$ and $1/t$ or :

$$Length \ of \ n/d = \ LCM(s-1, t-1)$$

I will just write "Length" from now on. Since s and t are primes not equal to 2 or 5, they are both divisible by 2. We have:

$$Length = \ LCM(s-1, t-1) \le (s-1)(t-1)/2 \qquad \text{(L)}$$

But since s and t are primes:

$$\varphi(d) = \varphi(s)\varphi(t) = (s-1)(t-1). \qquad \text{(P)}$$

Expression (L) is less than expression (P). So using there cannot be any full period composites for this case. The reader can work on generalizing this to other denominators.

Euler's Generalization of Fermat's Theorem gives only an upper bound on the length of the repeating pattern. What about the issue of immediate repeating? We can prove that all Case 1 proper fractions n/d will be immediate repeaters using the corollary to "Euler's Generalization of Fermat's Theorem and the corollary which followed it.

It is easy to see that $1/d$ must be an immediate repeater. The key fact is that there is a smallest power s such that

$$10^s = 1 (mod \ d)$$

So a complete list of remainders in the long division process of $1/d$ is:

$$1, 10(mod \ d), 10^2(mod \ d), ..., 10^{s-1}(mod \ d) \qquad (L1)$$

None of the these remainders after 1 equals 1. So no repetition of remainders is possible until the next remainder

$$10^s(mod \ d) = 1$$

This implies 1/d is an immediate repeater.

Consider the subfamily of the s fractions with the remainders in list (L1) divided by d. It is possible to show that these s fractions will have the same remainders and in the same cyclic order. All s fractions will be immediate repeaters and have the same digits with an s long digit pattern.

Here is an example to make things concrete. Consider the reduced fractions n/21. The remainders in the expansion of 1/21 are:

1,10,16,13,4,19 and then 1 starting the repetition.

The fraction 1/21 is a 6 digit repeater. If we take at random one of these remainders divided by 21, say 13/21. The remainders are now:

13,4,19,1,10,16 and then 13 starting the repetition.

We see that 13/21 has the same remainders in the same cyclic order as 1/21. The digits of 13/21 will therefore be the same as for 1/21 but start with a different digit and continue in the same cyclic order as 1/21.

What about other reduced fractions n/21? All reduced fractions n/21 have periods dictated by the Length of Period Theorem.. The smallest string of 9's divisible by both 3 and 7 is 999999—six nines. All reduced fractions n/21 have 6 digit repeating patterns. Euler's function gives the number of reduced fractions n/21:

$$\varphi(21) = \varphi(3)\varphi(7) = 2 * 6 = 12.$$

There are 12 such fractions. We have covered one subfamily with six fractions. Since each fraction n/21 has a 6 digit pattern, there can be only one other subfamily. The other subfamily (2/21, 5/21,7/21, 8/21, 11/21, and 20/21) has all the same characteristics of the first. The remainders and hence the digits are the same for each fraction and cyclic order rules. They also cannot fail to be immediate repeaters. If they were not immediate, they could not have 6 digit patterns.

The repeating pattern of reduced fractions n/d start immediately after the decimal place and its length is the lowest power of s such that

$$10^s = 1(mod\ d)$$

There is another way to tell how long the repeating pattern will be based on the primes that make up the factors of the denominator. We can use the Length of Period Theorem, LPT. The case of repeated prime factors was discussed in chapter 2. For extra practice take a look at this table for the case of repeated factors of 3.

Denominator	Size of Repeating Pattern
3	1
3^2	1
3^3	3
3^4	9
3^5	27
3^6	81
3^7	243

We see a jump at 3^3 from 1 to 3 in the repeating pattern. Why? The key is in the factorization of $(10^n - 1)$. The smallest number, $(10^n -1)$ divisible by 3^3 ,is $(10^3 -1)$. Hence the repeating factor is of size 3 by LPT.

The LPT method of determining the size of the repeating pattern is equivalent to finding the smallest s is such that

$$10^s = 1 (mod\ d) \qquad \text{(S)}$$

Because then

$$(10^s - 1)\ is\ divisable\ by\ d$$

In the specific case of d equal to 27, the smallest s is 3. Substituting into equation S:

$$10^3 = 1(mod\ 27)$$

Which implies that

$$(10^3 - 1)\ is\ divisable\ by\ 27$$

Since 3 is the lowest power of 10 for which this is true it means that all reduced fractions n/27 are three digit immediate repeaters.

See Table 2 at the end of the book.The fifth line of that table says that the size of the period of $1/3^5$ (or any reduced fraction $n/3^5$) is 27. This implies that the smallest n for which $(10^n -1)$ is divisible by 3^5 is 27! Here is the 27 digit decimal version of $1/3^5$, thanks to that 'mathssurrey' website:

.[004115226337448559670781893]

Note: I had to enter the equivalent, 1/243 to get this result. What other denominators have a 27 digit repeating pattern? It depends on the factorization of the number $(10^{27}-1)$. This number consists of 27 nines. Its prime factorization can be done on some websites and is

$3^5(37)(757)(9333667)(440334654777631)$

Fractions with these denominators are candidates for 27 digit repeaters. The reader is invited to determine which fractions are!

Suppose a table of the factorizations of (10^n-1) is not handy. We can still predict the size of the repeating pattern for some composite denominators. Let's look at some reduced fractions with denominators that are products of different primes in the table below. See if you can state the rule for the size of the repeating pattern which is given in the last column. Hint: Consider the size of the repeating patterns in the reciprocals of the individual primes in the middle column.

Pattern Size of Some Reduced Fractions with Composite Denominators

Denominator Factored into Two or More Primes	Respective Size of Repeating Patterns of the Reciprocals of the Two Primes in Factorization	Size of Repeating Pattern Of the denominator in Column One
3*11	{1,2}	2
3*37	{1,3}	3
11*37	{2,3}	6
11*7	{2,6}	6
11*41	{2,5}	10
37*41	{3,5}	15
11*7*41	{2,6,5}	30

Got it? The repeating pattern is the lowest common multiple of the numbers in column two. This fact is related to Length of Period Theorem in chapter two.

Let's move onto the Case 2 fractions where d has at least one factor of 2 and/or 5. For denominators with factors of 2 and/or 5 we will have a repeating pattern that starts sometime <u>after</u> the first decimal place.

Let's look at an example. Consider 1/6. The remainders (counting the numerator) in the long division process are:

$$1,4,4,4,4,\ldots$$

We see the remainder 4, not equal to the numerator, repeating forever. These remainders dictate the decimal expansion of 1/6. It is.1[6]. Thus the expansion of 1/6 is a one digit repeater, but the repeating digit starts <u>after</u> the first digit. The conditions of Euler's generalization theorem are seen to be important. Six and ten are not relatively prime. The theorem does not guarantee that

$$10^{\varphi(6)} = 1(mod\ 6)$$

And in fact no positive power of 10 gives 1 mod 6. Every positive power of 10 gives 4 mod 6.

We have the remainder 4 repeating and it is not equal to the numerator of the fraction. This only happens when there is a factor of 2 and/or 5 in the denominator! It never occurs with prime or with composite denominators having no factors of 2 or 5.

The last question about Case 2 fractions is, " How many decimal places will it be before the repeating pattern starts?" We will define the number of places before the repeating starts as, "initial places". Obviously for each factor of 10 in the denominator a leading zero is generated. Factors of 2 and/or 5 alone create one initial place. Compare having 2 or 5 to cases without them:

Original Fraction	Decimal Expansion Without the 2's or 5's	Expansion of Original Fraction
$1/(2*3)$	$1/3=[.3]$	.1[6]
$1/(2^2*3)$	$1/3==[.3]$	.08[3]
$29/(5^1*37)$	$29/37=[.78]$	.1[567]
$29/(5^2*37)$	$29/37=[.783]$	.03[135]
$29/(2^3*5^2*37)$	$29/37==[.783]$	.003[918]
$29/(2^2*5^5*37)$	$29/37==[.783]$	.00006[270]

The pattern is pretty clear. Each power of 2 or of 5 alone adds a digit before the repeating pattern starts. Note: We are requiring that n/d is a reduced fraction. For example, 38/99 and 38/(2*99) are both immediate repeaters but the latter is not a reduced fraction.

What if 2's and 5's are both present? We note the highest common power, c. The common power c causes c zeros before the repeating pattern. Finally, the repeating pattern will start in additional places equal to the number of left over powers greater than c. Consider the last row of the table with denominator (2^2*5^5). The highest common exponent that two and five have is two. This creates two zeros before the repeating pattern. There are three powers of 5 left over. So there will be three plus two equals five initial places before the repeating pattern starts, as seen in the last column.

Table 4 at the end of the book gives many examples with reasons for the number of initial places and the number of repeating places. The exercises which follow have more challenges.

Chapter 5 Exercises

For the fractions in exercises 1 through 3 do the following:

a) Factor the denominator
b) State the number of initial digits if any before the repeating process starts.
c) State the number of repeating places and give a reason:

1. $\dfrac{23}{407}$

2. $\dfrac{17}{2214}$

3. $\dfrac{97}{348940}$

4.) This exercise will contrast the clocks of n/37 with those of the composite denominator fractions n/27. Recall in exercise 3 of chapter 3 that the family n/37 are three digit repeaters and the whole family of 36 fractions can be organized into (36)/3 equals 12 clocks. The sum of the fractions in each clock adds to an integer.

Consider the family of fractions with the composite denominator 27.:

$$1/27, 2/27, 3/37, \ldots, 26/27$$

a) Only the non reducible fractions are three digits repeaters. All of these are three digit repeaters like the prime family n/37. How many three digit repeaters are there and how many clocks can be organized to display all the fractions that are three digit repeaters?

b) In chapter 3 we showed that the sum of the fractions in any clock with a prime denominator is an integer. Does the sum of any clock in the non-reducible fractions of n/27 sum to an integer? How about the sum of all the non- reducible fractions in all the clocks?

For exercises 5 through . Study Table 2 and the explanation with it. We can use the table to create fractions with desired properties.

5. What is the smallest denominator that has the <u>prime</u> lengths:

a) a three decimal repeating pattern?

b) a five decimal repeating pattern?

c) a seven decimal repeating pattern?

6. What is the smallest denominator that has the <u>composite</u> lengths:

a) a four decimal repeating pattern?

b) a nine decimal repeating pattern?

c) a 10 decimal repeating pattern?

7. Predict the period of the following fractions:

a) $1/(9*37)$ b) $7/(41*27)$ c) $8/(13*271)$

8. What is largest reciprocal that has the following properties:

a) 2 initial places, 2 repeating digits?

b) 3 initial places, 4 repeating digits?

c) 5 initial places, 7 repeating digits?

6 Converting Infinite Series to Fractions

Recall the infinite geometric sum formula from chapter 1. .

$$1 + r + r^2 + \ldots = \frac{1}{1-r} \quad for \quad -1 < r < 1 \quad (1)$$

We can use this formula to convert a two digit repeating decimal to a fraction:

$$.[54] = \frac{54}{100} + \frac{54}{100^2} + \ldots$$

$$= \frac{54}{100}\left(1 + \frac{1}{100} + \frac{1}{100^2} + \ldots\right) = \frac{54}{100} \, \frac{1}{(1 - {}^1/_{100})}$$

$$= \frac{54}{100} \, \frac{1}{99/_{100}} = \frac{54 \; 100}{100 \; 99}$$

$$= \frac{54}{99} = \frac{6}{11}$$

The same thing can be done for repeating decimals with any number of repeating places.

My last example, patient reader, shows two ways to convert the three digit repeating decimal [.296] to a reduced fraction:

$$[.296] = \frac{296}{1000} + \frac{296}{1000^2} + \ldots$$

$$= \frac{296}{1000}\left(1 + \frac{1}{1000} + \frac{1}{1000^2} + \ldots\right)$$

$$= \left(\frac{1}{1 - \frac{1}{1000}}\right) \frac{296}{1000}$$

$$= \frac{296}{1000} \quad \frac{1}{999/1000}$$

$$= \frac{296}{999} = \frac{8 * 37}{27 * 37} = \frac{8}{27}$$

The same thing can be done by realizing that:

$$[.296] = \frac{296}{999} = \frac{8 * 37}{27 * 37} = \frac{8}{27}$$

In that 'mathssurrey' website one can also type in a repeating pattern and the site will output the reduced fraction.

As a parting challenge, the reader is asked to show that the base 33 form of 1/17 given in chapter 3 as [1 31] is correct using an infinite series.

Chapter 6 Exercises

1. All fractions that are two digit repeaters can be characterized as the family with denominator x in the list:

$$1/x, 2/x, 3/x, \ldots, 98/x$$

Note this includes repeated digits such as .[66] but excludes .[99] because it equals one.

a) What is x ? How many such two digit repeaters are there?

b) Of the 98 such two digit repeaters in part (a), how many have **different digits?**

c) Of the 90 two digit repeaters with different digits how many are:

i) divisible by 3?

ii) divisible by 9?

iii) divisible by 11?

2. This is another way to look at the 98 fractions of exercise 1. Of the 98 fractions

1/99,2/99, 3/99,...,98/99

a) How many cannot be reduced?
b)How many can be reduced?
c) Of the 38 that can be reduced how many are :
   i) divisible by 3?
   ii) divisible by 11?
  iii) divisible by 3 and 11?

3. Consider the repeating 6 place decimal .[360360].

a) What is its fraction form with denominator $(10^n - 1)$? Write the specific n.

b) Is the fraction in part (a) reduced? If not what are the possible common divisors based on the denominator? Note: Table 1 at the end of the book lists the factors of $(10^n -1)$ for many n.

c) What are the common divisors of 360360 and 999999? What is the reduced fraction version of 360360/999999?

4.) Of the 999,998 fractions which have 6 repeating digits:

1/999999,2/999999,...,999998/999999

How many are irreducible?

Answers and Solutions to Exercises

Chapter 1

1.) The reciprocals of powers of two: 2,4,8,16,32,and 64,.
The reciprocals of powers of five: 5,and 25.
Combinations of reciprocals with factors of 2 and 5:
10,20,40 50 , 80, and 100.

2.) We can look first at the reciprocals of primes except for 2 and 5:
There are 24; the reciprocals of :

3,7,11,13,17,19,23,29,31,37,41,43,47,53,59,61,67,71,73,79,83,89,97,and
101.
There are also 16 composite denominators that are immediate
repeaters. The reciprocals of:

9,21,27,33,39,49,51,57,63,69,77,81,87,91,93, and 99.

In all there are 40 immediate repeaters .

3.) The reciprocal of the prime 97 has a 96 period repeating pattern!
When the reciprocal of a prime p has the longest possible period of (p-1) it is called a long prime. See chapter 2.

4. a) 1/37 b) 1/101 c) 1/41

Chapter 2
1. All these fractions have 5 digit repeating patterns because 41 divides
into 99999 and no smaller string of nines. The expansions are a)
.[03717], b) .2[31707], c) .011[58536]
2. a) One, 3^2 divides into 9
b) Three, 3^3 divides into 999
c) Nine, 3^4 divides into 999999999
d) Eighteen , Both 3^4 and 11 divide into $(10^{18} -1)$ AND NO
SMALLER STRING OF NINES!
3. a) Nothing

b) The 2's and 5's delayed the onset of the decimal pattern. In chapter 5 we will see a rule to predict how long the the decimal pattern is delayed.

Chapter 3

1.) 10,26, and 1. These are exactly the remainders in the long division process of 1 divided by 37. The last modular value of 1 is crucial. It equals the numerator of the fraction and also implies that the remainders will now repeat the pattern: 10,26,1 forever!

2.) The appropriate Mod values are:
$$10 \times 1 (Mod\ 41) = 10,$$
$$10^2 \times 1 (Mod41) = 8$$
$$10^3 \times 1, (Mod41) = 16$$
$$10^4 \times 1 (Mod\ 41) = 37,$$
$$and\ 10^5 \times 1 (Mod41) = 1$$
The process will now repeat. Thus we know that 1/41 is a 5 digit repeater.

3) Here I write the Mod symbol around the whole operation:
$Mod(10*52,101)=15$
$Mod(10^2 *52,101)=49$
$Mod(10^3 *52,101)=86$
$Mod(10^4 *52,101)=52$, We are back to the numerator. Thus   52/101 is a 4 digit repeater. In fact, all 100 fractions n/101 are 4 digit repeaters.

4) $1/10= .0[0011]_2$, $1/10= .[0022]_3$, $1/10= .0[12]_4$,
$1/10= .0[2]_5$, $1/10= .[0462]_7$,

5.) Bases 3 and 7 were immediate repeaters. The reason is that only 3 and 7 were relatively prime to 10.

6 a) Only bases 5,7,11,13 and 17
 b) All the other bases except 6 and 12.
c) 6 and 12.

7). The last column should be:
  7 divides into $(2^3 -1)$
  7 divides into $(3^6 -1)$
  7 divides into $(4^3 -1)$
  7 divides into $(5^6 -1)$
  7 divides into $(6^2 -1)$

7 divides into $(8^1 - 1)$
7 divides into $(9^3 - 1)$

Chapter 4

1. The arithmetic series of $1 + 2 + \dots (p-1)$ sums to $p(p-1)/2$. The sum of the fractions is $(p-1)/2$, a whole number for any odd p.

2. a) 27 and 72.
   b) From problem 1 we know that the sum of the fractions

$$1/p + 2/p + 3/p +, \dots, + (p-1)/p$$

is an integer. The decimal version of these fractions must sum to an integer. All fractions in n/7 have the same digits. It is sufficient to look at the sum of the digits in just 1/7 to see how the 6 fractions will add to an integer. (Same for n/17) To get an integer the sum must be of the form:

$$(Integer).9999\dots = Integer + 1$$

Note: Nine repeated forever equals one. To get an integer, the sum of the digits must be 9 at every decimal place. When all (p-1) fractions are added the sum of the digits could be 9 or if greater than 9 then sum of the digits in the sum must be 9! We saw in part (a) two examples of this. For 1/7 the sum of the digits was 27, but its sum is (2+7=9). In 1/17 the sum was 72, sum these again to 9. Other possible two digit sums are 18,81,36, 63, 45 and 54..

   c) The sum of the digits in 126 is 9. The decimal sum of all the fractions in the family will be of the form:

$$Integer.[9999] = Integer + 1$$

This is the decimal way of ensuring that the sum of all the fractions in the family n/29 is an integer.

3. a) Three digits b) 36/3 implies 12 clocks
   c) One or two

Chapter 5
1. a) $407 = (11)(37)$
   b) No initial places because no factors of 2 or 5.

c) The primes 11 and 37 have repeating pattern of lengths 2 and 3 respectively. So the fraction will have a repeating pattern of the LCD of these numbers, which is 6 digits long.

2. a) $2214 = 2*3^3 *41$

    b) One initial place of zero because of the factor 2.

    c) 3^3 and 41 have repeating patterns of length 3 and 5 respectively so the fraction has a repeating pattern of length 15.

3. a) $348940=2^2 *5 *73*239$

    b) There are 2 initial places—2 Times 5 causes one place and then one more factor of 2 causes an additional initial place.

c) 73 and 239 have patterns of 8 and 7 respectively. The LCD of these numbers is 56, so this fraction is a 56 digit repeater.

4.a) (26-8) implies 18 three digit repeaters in this family. There will be three fractions in each clock so 18/3 makes 6 clocks of three digit repeaters.

b) No, but the sum of all the fractions in the six clocks is the integer 6.

5a) 3^3 b) 41,c)239

6a) 101,b) 3^4 c)11*41

7 a) 3 b) 15 c)30

8 a) 1/(4*11) b)1/(8*101) c)1/(32*239)

Chapter 6

1a) 99, 98

b) 90

ci) 30, cii)10, ciii)none

2a) . They are the ones with numerators relatively prime to 99. The number of them can be found easily using the properties of Euler's 'Φ' function. $\Phi(99)=60$.

    b) 98-60=38

    ci)32 ii)8, iii)2

3a) $\dfrac{360360}{10^6 - 1}$

or $\dfrac{360360}{999999}$

b) No. Based on the denominator the possible common divisors are 3 to the power 1,2, and also 3,7,11,13,37.

c) Common factors are

$3^2,7,11$ and 13.

The reduced fraction is 40/111.

4. Use Euler's Φ function:

$$\Phi (999999)= \Phi(9*7*11*13)$$
$$= \Phi(9)\ \Phi(7)\ \Phi(11)\ \Phi(13)$$
$$= 6*6*10*12$$
$$=155,520 \text{ fractions}$$

Tables

1. Length of Non-Terminating Prime Denominators

Prime p	Decimal Length of Repeating Pattern Fraction n/p	Primes That Shares the Same Length
3	1	None
7	6	{13}
11	2	None
13	6	(7)
17	16	(5882353)
19	18	$(10^{19} - 1)/9))$
23	22	$(10^{23} - 1)/9))$
29	28	(281),(121499449)
31	15	(2906161)
37	3	None
41	5	(271)
47	46	(139),(2531),(549797184491917)
53	13	(79),(265371653)
59	58	(154083204930662557781201849)
61	60	(4188901),(39526741)
67	33	(1344628210313298373)
71	35	(123551),(102598800232111471)
73	8	(137)
79	13	(53),(265371653)
83	41	(1231),538987), (201763709900322803748657942361)
89	44	(10566892961)
97	96	?

57

1. Any prime that has a n-length repeating pattern divides   evenly into $(10^n -1)$. The number $(10^n -1)$ is an integer equal to n nine's. The next table has the factorizations of
$(10^n -1)$ for n equal one to ten.

Example: The six fractions: n/7 for n=1,2,3...,6 are all 6 digit repeaters. The next table shows that:

$$(10^6 - 1) = 3^3 \cdot 7 \cdot 11 \cdot 13 \cdot 37$$

The significance of this factorization is that denominators with these factors have repeating patterns that are either:

(a) 6 digit repeating pattern (13 is the only one besides 7)

or

 (b) have a decimal repeating pattern length that divides into 6: We verify as follows:

Denomonators of 3^3 *and 37 have length* 3.

Denominators of 11 *has length* 2.

2. Factorizations of $10^n - 1$

NOTE: The bold face factors indicate their first occurrence in the table. Fractions with bold face denominators have repeating patterns of length n, equal to row they are in.

$10^n - 1 = \ldots$	PRIME FACTORIZATION
$10^1 - 1 = 9$	3^2
$10^2 - 1 = 99$	$(3^2)\mathbf{11}$
$10^3 - 1 = 999$	$\mathbf{(3^3)(37)}$
$10^4 - 1 = 9999$	$(3^2)(11)\mathbf{(101)}$
$10^5 - 1 = 99999$	$(3^2)\mathbf{(41)(271)}$
$10^6 - 1 = 999999$	$(3^3)\mathbf{(7)}(11)\mathbf{(13)}(37)$
$10^7 - 1 = 9999999$	$(3^2)\mathbf{(239)(4649)}$
$10^8 - 1 = 99999999$	$(3^2)(11)\mathbf{(73)}(101)\mathbf{(137)}$
$10^9 - 1 = 999999999$	$\mathbf{(3^4)}(37)\mathbf{(333667)}$
$10^{10} - 1 = 999999999$	$(3^2)(11)(41)(271)\mathbf{(9091)}$

How does this table apply to fractions? Consider the case for n equal to 6. From the table:

$$10^6 - 1 = 999999 = 3^3 \cdot 7 \cdot 11 \cdot 13 \cdot 37$$

This means that any fractions with denominators 3, 3^2, 3^3, 7, 11, 13 and 37 will have repeating patterns that are either 6 digits long or a length which divides into 6. More explicitly, if n is any integer in a reduced fraction then:

n/3 has repeating pattern of length 1.
$n/3^2$ has repeating pattern of length 1.
$n/3^3$ has repeating pattern of length 3.
n/7 has repeating pattern of length 6.
n/13 has repeating pattern of length 6, and
n/37 has repeating pattern of length 3.

Denominators with combinations of these integers will have repeating patterns whose length is the lowest common multiple of the lengths of the repeating patterns for the reciprocals of the individual prime factors.

Finally, since 7 and 13 are in bold face, fractions with these denominators have decimal pattern of length 6. Those not in bold face have smaller lengths which divide into six.

Some prime denominators have repeating lengths r which are themselves prime. Examples are 37,41 and 53. See Table 1. In these cases the factorization of $(10^r - 1)$ gives primes whose repeating length is either the prime r or equals one. For example the number five is a prime and

$$10^5 - 1 = 3^2 \cdot 41 \cdot 271$$

Fractions with these denominators have length 5, (41 and 271) or have length one (3^2).

3. Lengths and Number of Subfamilies of Some Prime Denominators

n/p,for p a Prime and
n = 1,2,3,...,(p-1)

Fraction Family	Number of Digits s in Repeating Pattern	Number of Subfamilies or Clocks = (p-1)/s
n/7	6	1
n/11	2	5
n/17	16	1
n/19	18	1
n/29	28	1
n/31	15	2
n/37	3	12
n/41	5	8
n/101	4	25
n/137	8	17
n/9091	10	909
n/333667	9	37,074

An example using this table:

The 40 fractions in the family n/41 all are 5 digit repeaters. For Instance,

1/41 = [.02439].

There are (41-1)/5= 8 subfamilies in all. 1/41 is a member of one subfamily, all of which uses the digits 0,2,4,3 and 9. Another member of this subfamily is

$\frac{10}{41}$ = [.24390].

The reader can find the other 3 members which use these 5 digits. Also the reader can make a clock for this subfamily, as explained in chapter 4. While you're at it, make 7 more clocks for the other 7 subfamilies in n/41!

4. Some Composite Denominator Expansions

Explanation: <u>Initial Places</u> is the number of digits before the repeating pattern starts. In the third column is the reason for the number of Initial Places followed by the reason for the number of repeating places. Recall from Chapter 5 that each power of 2 or 5 in the factorization of the denominator causes an Initial Place in the decimal. For example: $52= (2^2)(13)$. Thus there 2 Initial places in 1/52. The number of digits in the repeating pattern is determined by the remaining factors (excluding 2 and 5) . The number of digits in the repeating pattern is n, the <u>smallest</u> power of $(10^n -1)$ which the remaining factors divide into. Continuing with 1/52, 13 divides into $10^6 - 1$, which is the reason there is a repeating pattern that is 6 digits long.

Fraction	Initial Places-- Repeating Places	Reasons For: Initial--Repeating Places Based on Factors of the Denominator
$1/(2^2*3)$	2--1	Factor of 2^2 –Factor 3 divides into 9
$1/(3*5)$	1--1	Factor of 5.-- Factor 3 divides into 9
$1/3^3$	0--3	No powers of 2 or 5—27 divides into 10^3-1
$1/(10*3)$	1--1	Factor of 10 creates one initial place—factor 3 divides into (10^1-1)
$1/52=1/((2^2)(13))$	2--6	Factor of 2^2--Factor 13 divides into 10^6-1
$1/(2^5)(3)$	5--1	Factor of 2^5—Factor 3 divides into 9
$1/(3^2*41)$	0--5	No factors of 2 or 5—Both 3^2 and 41 divide into $(10^5 -1)$
$1/(3^3*41)$	0--15	No Factors of 2 or 5--3^3 and 41 divide into $(10^{15}-1)$
$1/(3^4*41)$	0--45	No Factors of 2 or 5—3^4 and 41 divide into $(10^{45}-1)$

www.ingramcontent.com/pod-product-compliance
Lightning Source LLC
Chambersburg PA
CBHW070919180526
45168CB00005B/2069